Level 1 Practice Problems

Sasha's World™

Number Line Equations

Fun Algebra

Princess Sasha Saves Baby Dinosaurs

by Courtney West, Ph.D.

For my daughter, Jazzmine C., with love

To my parents, Randall and Audrey Gaines, with love

To my siblings, Dawn W., Crystal G., and Randy Jr. G., with love

To my friend, Clay Holleman, with love

12-Book Algebra 1 Series

Book 1: Princess Sasha Saves Baby Dinosaurs: Fun Algebra Level 1 Practice Problems

ISBN-10: 1548665363
ISBN-13: 978-1548665364

Table of Contents

Author's Preview

Congratulations on reading *Princess Sasha Saves Baby Dinosaurs: Fun Algebra* to your child. He/she is now ready to solve Level 1 **_Number Line_** Algebra equations. By solving the **45** equations presented in this Practice Problems book, your child will strengthen skills he/she developed from the above title. The above fairy tale about baby dinosaurs is the first release of a 12-book Algebra 1 series for children ages 4 and older.

Princess Sasha Saves Baby Dinosaurs uses fun and engaging puzzles and tricks to show your child how to solve Algebra equations such as
X + 3 = 4   *and*   **X - 2 = 1**   *and*   **X – (- 2) = 3**.

Princess Sasha Algebra books are interactive and enable parents (and others) to teach basic Algebra to preschool students. This is true even if: 1) the children don't know how to read and 2) the parent/reader has no understanding of Algebra. For each Practice Problem below, cover everything except the Algebra **_equation_**. Then you or your child should **_draw_** this Number Line:

-10 -9 -8 -7 -6 -5 -4 -3 -2 -1 0 1 2 3 4 5 6 7 8 9 10

If your child is using a pencil with an eraser, he/she can use the same **_Number Line_** for the 45 equations. For each Practice Problem, your child *should <u>draw</u>: 1) the <u>**equation**</u>, 2) <u>**large dots**</u>, 3) an arrow in the right direction, and 4) the <u>**solved**</u> equation. Then uncover the page to check the accuracy of his/her **Number Line** and solution. Below are three examples of the type of Number Line Algebra equations presented in this Practice Problems book for Princess Sasha Saves Baby Dinosaurs: Fun Algebra.*

1. X + 4 = 6

-10 -9 -8 -7 -6 -5 -4 -3 -2 -1 0 1 2 3 4 5 6 7 8 9 10

$$X = 2 \longrightarrow 2 + 4 = 6$$

2. X + (- 4) = 6

-10 -9 -8 -7 -6 -5 -4 -3 -2 -1 0 1 2 3 4 5 6 7 8 9 10

$$X = 10 \longrightarrow 10 + (- 4) = 6$$

3. X - (- 4) = 0

-10 -9 -8 -7 -6 -5 -4 -3 -2 -1 0 1 2 3 4 5 6 7 8 9 10

$$X = - 4 \longrightarrow - 4 - (- 4) = 0 \longrightarrow - 4 + 4 = 0$$

Princess Sasha Saves Baby Dinosaurs

Algebra Practice Problems – <u>Pictures</u>

<u>Number Line Equations</u>

$$X + 2 = 5$$

X = 🦋 🦋 🦋

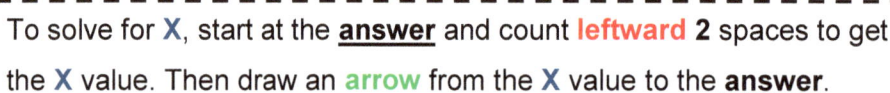

```
-10 -9 -8 -7 -6 -5 -4 -3 -2 -1  0  1  2  3  4  5  6  7  8  9  10
```

To solve for **X**, start at the **<u>answer</u>** and count **leftward** 2 spaces to get the **X** value. Then draw an **arrow** from the **X** value to the **<u>answer</u>**.

Final Answer:
X = 3

Check Your Answer:
X + 2 = 5
3 + 2 = 5 ✓

Algebra Practice Problems – <u>Pictures</u>

<u>Number Line Equations</u>

$$X + 6 = 9$$

$X =$

-10 -9 -8 -7 -6 -5 -4 -3 -2 -1 0 1 2 3 4 5 6 7 8 9 10

To solve for **X**, start at the **<u>answer</u>** and count **leftward 6** spaces to get the **X** value. Then draw an **arrow** from the **X** value to the **<u>answer</u>**.

Final Answer:

X = 3

Check Your Answer:

X + 6 = 9

3 + 6 = 9

Algebra Practice Problems – <u>Pictures</u>

<u>Number Line Equations</u>

$$X + 8 = 9$$

X =

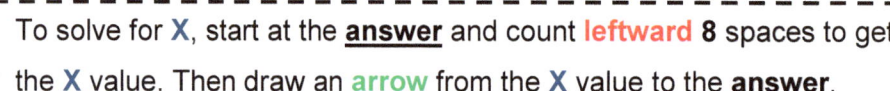

-10 -9 -8 -7 -6 -5 -4 -3 -2 -1 0 1 2 3 4 5 6 7 8 9 10

> To solve for **X**, start at the **<u>answer</u>** and count **leftward** 8 spaces to get the **X** value. Then draw an **arrow** from the **X** value to the **<u>answer</u>**.

Final Answer:
X = 1

Check Your Answer:
X + 8 = 9
1 + 8 = 9 ✓

Algebra Practice Problems – <u>Pictures</u>

<u>Number Line Equations</u>

$$X - 1 = 3 \quad \textit{same as} \quad X + (-1) = 3$$

X =

-10 -9 -8 -7 -6 -5 -4 -3 -2 -1 *0* 1 2 3 4 5 6 7 8 9 10

> To solve for **X**, start at the **answer** and count rightward **1** space to get the **X** value. Then draw an **arrow** from the **X** value to the **answer**.

Final Answer:

X = 4

Check Your Answer:

$$X + (-1) = 3$$
$$4 + (-1) = 3 \checkmark$$

Algebra Practice Problems – <u>Pictures</u>

<u>Number Line Equations</u>

X - 2 = 3   *same as*   **X + (-2) = 3**

X =

-10 -9 -8 -7 -6 -5 -4 -3 -2 -1 0 1 2 3 4 5 6 7 8 9 10

To solve for **X**, start at the **<u>answer</u>** and count **rightward** **2** spaces to get the **X** value. Then draw an **arrow** from the **X** value to the **<u>answer</u>**.

Final Answer:

X = 5

Check Your Answer:

X + (-2) = 3
5 + (-2) = 3 ✓

Algebra Practice Problems – *Pictures*

Number Line Equations

$$X + (-2) = -6$$

$X = \ ^{-}$

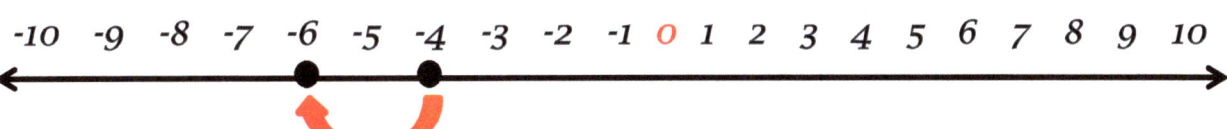

| -10 | -9 | -8 | -7 | -6 | -5 | -4 | -3 | -2 | -1 | 0 | 1 | 2 | 3 | 4 | 5 | 6 | 7 | 8 | 9 | 10 |

To solve for **X**, start at the **answer** and count rightward 2 spaces to get the **X** value. Then draw an **arrow** from the **X** value to the **answer**.

Final Answer:

$X = -4$

Check Your Answer:

$$X + (-2) = -6$$
$$-4 + (-2) = -6 \ \checkmark$$

8

Algebra Practice Problems – <u>Pictures</u>

<u>Number Line Equations</u>

$$X + (-2) = -5$$

$$-10 \quad -9 \quad -8 \quad -7 \quad -6 \quad -5 \quad -4 \quad -3 \quad -2 \quad -1 \quad 0 \quad 1 \quad 2 \quad 3 \quad 4 \quad 5 \quad 6 \quad 7 \quad 8 \quad 9 \quad 10$$

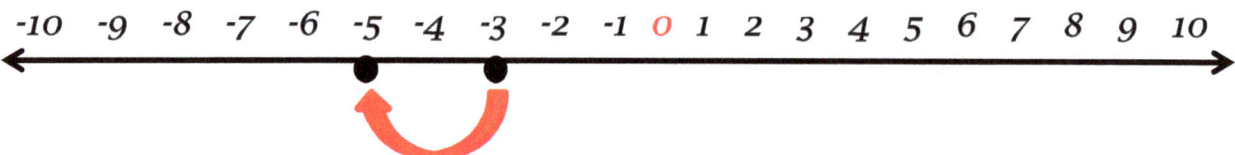

To solve for **X**, start at the **answer** and count **rightward** 2 spaces to get the **X** value. Then draw an **arrow** from the **X** value to the **answer**.

Final Answer:
X = - 3

Check Your Answer:
$X + (-2) = -5$
$-3 + (-2) = -5 \checkmark$

9

Algebra Practice Problems – <u>Pictures</u>

<u>Number Line Equations</u>

X + 1 = 7

X =

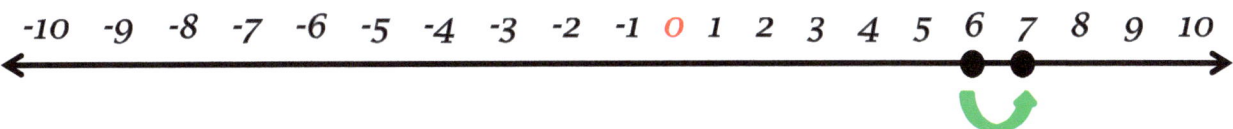

To solve for **X**, start at the **<u>answer</u>** and count **leftward** **1** space to get the **X** value. Then draw an **arrow** from the **X** value to the **<u>answer</u>**.

Final Answer:

X = 6

Check Your Answer:

X + 1 = 7

6 + 1 = 7 ✓

10

Algebra Practice Problems – <u>Pictures</u>

<u>Number Line Equations</u>

$$X + (-3) = -3$$

$$X =$$

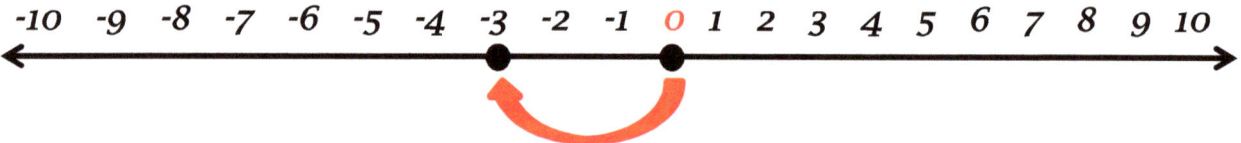

-10 -9 -8 -7 -6 -5 -4 -3 -2 -1 0 1 2 3 4 5 6 7 8 9 10

> To solve for **X**, start at the <u>**answer**</u> and count **rightward** 3 spaces to get the **X** value. Then draw an **arrow** from the **X** value to the <u>**answer**</u>.

Final Answer:
X = 0

Check Your Answer:
X + (-3) = -3
0 + (-3) = -3 ✓

Algebra Practice Problems – Pictures

Number Line Equations

$$X - 9 = -4 \quad \textit{same as} \quad X + (-9) = -4$$

X =

To solve for **X**, start at the **answer** and count **rightward** 9 spaces to get the X value. Then draw an **arrow** from the X value to the **answer**.

Final Answer:

X = 5

Check Your Answer:

$$X + (-9) = -4$$
$$5 + (-9) = -4 \checkmark$$

 # The Number Line Game

-10 -9 -8 -7 -6 -5 -4 -3 -2 -1 0 1 2 3 4 5 6 7 8 9 10

Right now, your child can use his/her knowledge of Number Lines to figure out who won **The Number Line Game**: Jazmin, José, Joshua or Sasha? First, show your child the game board on p. 14 and give him/her one or more crayons, along with a blank sheet of paper. Then tell him/her to do the **4** things below to figure out who won **The Number Line Game**. See p. 15 for the answer.

Steps to Figure out the Winner:

- Draw **4** **Number Lines** like the one below. Draw each player's name and/or assigned number above one Number Line (i.e. "1" for Jazmin, "2" for Sasha, "3" for José and "4" for Joshua).

  -10 -9 -8 -7 -6 -5 -4 -3 -2 -1 0 1 2 3 4 5 6 7 8 9 10

- Under the Number Line, <u>draw</u> its **equation** (e.g. 10 + **(-4)** = 6 for Sasha). Use "10" as the **X** value (i.e. number of animal cards for each player). For Sasha, "**-4**" is the **known number** (i.e. number of crossed-out cards). And "6" is the _answer_ (i.e. 10 cards minus 4 cards equal _6 cards_).

- On Sasha's Number Line, draw a **large dot** under _answer_ "6." From the answer, count spaces representing **known number** "-4." So count 4 spaces _rightward_ to "10" (i.e. the **X** value), and draw a **large dot** under it.

13

- Draw <u>an arrow</u> from "10" to the answer. The player with the <u>biggest answer</u> won the game. If your child doesn't say "Joshua won," ask him/her to recount the cards for each player and change the dots on the <u>**Number Lines**</u> and the numbers in the <u>**Algebra equations**</u> he/she drew.

Winner of The Number Line Game: Joshua!

Jazmin - 1

-10 -9 -8 -7 -6 -5 -4 -3 -2 -1 0 1 2 3 4 5 6 7 8 9 10

10 + (- 5) = 5

Sasha - 2

-10 -9 -8 -7 -6 -5 -4 -3 -2 -1 0 1 2 3 4 5 6 7 8 9 10

10 + (- 4) = 6

José - 3

-10 -9 -8 -7 -6 -5 -4 -3 -2 -1 0 1 2 3 4 5 6 7 8 9 10

10 + (- 6) = 4

Joshua - 4

-10 -9 -8 -7 -6 -5 -4 -3 -2 -1 0 1 2 3 4 5 6 7 8 9 10

10 + (- 2) = 8

Parents can create ***The Number Line Game*** using computer clip art, low-cost materials from retail stores, and the game rules below.

Game Rules

- Each of **4 players** takes **10 animal cards** and places them in the squares in front of him/her. Each card is worth **1** point. Be sure to shuffle the cards.

- Each player draws a **Number Line** and **equation**. The **X** value "**10**" is for the initial number of **animal cards**. The crossed-out cards represent the **known number**, and the *answer* is 10 <u>minus</u> the known number.

- Each player places one herbivore animal card in the **Card Box** for each **carnivore** among the *first 10 cards* he/she selects. Carnivores eat herbivores so players **lose 1 herbivore card** for each carnivore card.
    a. ***Carnivores*** are shown as baby **dinosaurs**, **lions** and **tigers**.
    b. ***Herbivores*** eat plants so no points are lost for holding cards with a picture of an **elephant**, **zebra**, **goat**, **duck**, **butterfly**, **bee**, **puppy** or **bird**. <u>Note:</u> wild dogs eat animals and birds usually eat insects.

- <u>One player at a time</u> takes **one** Number Card. Thus, throughout the game, each child takes <u>20 number cards</u> with numbers **0, 1, -1, 2** and **-2.** If the selected card has a **negative number** (i.e. **-1** or **-2**), he or she immediately must place the corresponding number of **animal cards** in the **Card Box**.

- Cards with **positive numbers** (i.e. **1** or **2**) allow players to retrieve corresponding numbers of **animal cards** from the <u>Card Box</u>. A "**0**" card means <u>no card is lost or received</u>. All number cards selected: game ends.

Princess Sasha Saves Baby Dinosaurs

Algebra Practice Problems – <u>Large Dots</u>

X + 3 = 6

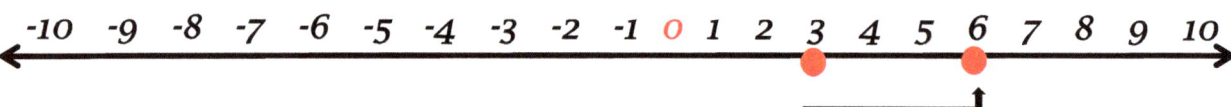

> To get the **X** value, start at the **<u>answer</u>** and count **leftward** when the **known number** is <u>positive</u> and **rightward** when it's <u>negative</u>. Then draw an **arrow** from the **X** value to the **<u>answer</u>**.

X + (-3) = 6

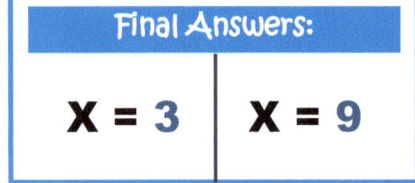

Final Answers:	
X = 3	**X = 9**

Check Your Answers:	
X + 3 = 6 **3 + 3 = 6** ✓	**X + (-3) = 6** **9 + (-3) = 6** ✓

Algebra Practice Problems – <u>Large Dots</u>

X + 4 = 6

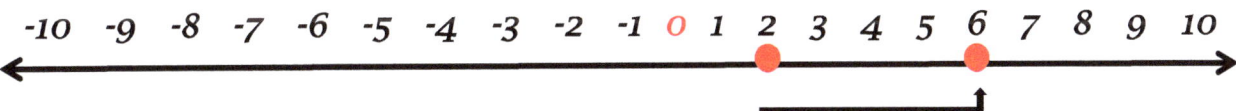

To get the **X** value, start at the **<u>answer</u>** and count **leftward** when the **known number** is <u>positive</u> and **rightward** when it's <u>negative</u>. Then draw an **arrow** from the **X** value to the **<u>answer</u>**.

X + (- 4) = 6

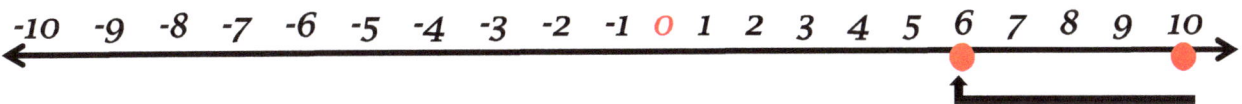

Final Answers:	
X = 2	X = 10

Check Your Answers:	
X + 4 = 6	X + (-4) = 6
2 + 4 = 6 ✓	10 + (-4) = 6 ✓

18

Algebra Practice Problems – <u>Large Dots</u>

X + 1 = 4

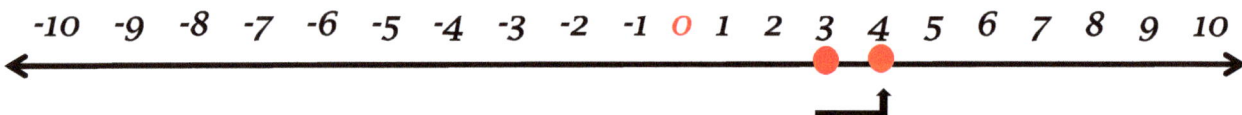

-10 -9 -8 -7 -6 -5 -4 -3 -2 -1 *0* 1 2 3 4 5 6 7 8 9 10

> To get the **X** value, start at the **answer** and count **leftward** when the **known number** is <u>positive</u> and **rightward** when it's <u>negative</u>. Then draw an **arrow** from the **X** value to the **answer**.

X + (- 1) = 4

-10 -9 -8 -7 -6 -5 -4 -3 -2 -1 *0* 1 2 3 4 5 6 7 8 9 10

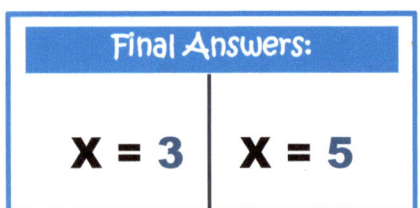

Final Answers:	
X = 3	X = 5

Check Your Answers:	
X + 1 = 4	X + (-1) = 4
3 + 1 = 4 ✓	5 + (-1) = 4 ✓

Algebra Practice Problems – Large Dots

X + 6 = 8

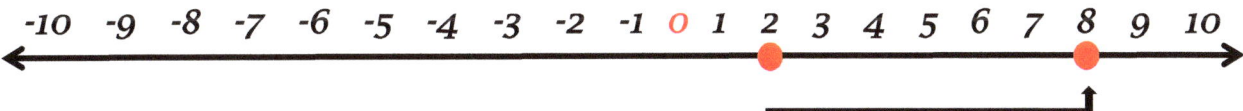

```
To get the X value, start at the answer and count leftward when the known number is positive
and rightward when it's negative. Then draw an arrow from the X value to the answer.
```

X + (- 10) = - 8

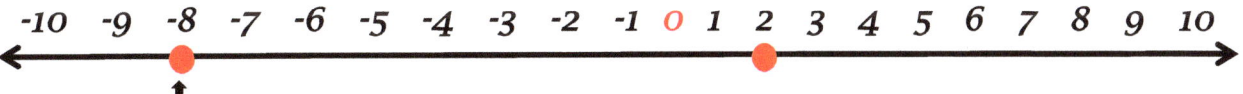

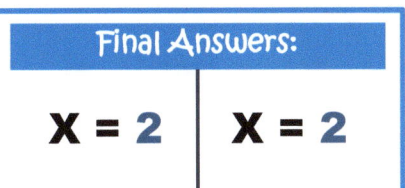

Final Answers:	
X = 2	X = 2

Check Your Answers:	
X + 6 = 8 2 + 6 = 8 ✓	X + (-10) = -8 2 + (-10) = -8 ✓

Algebra Practice Problems – <u>Large Dots</u>

X + 8 = 5

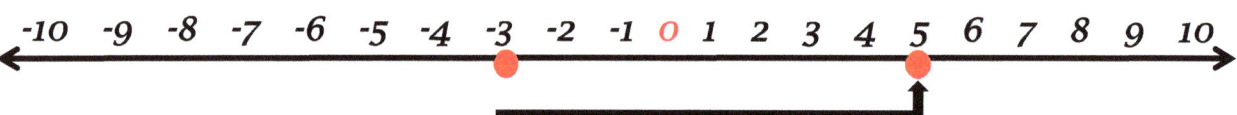

> To get the **X** value, start at the <u>**answer**</u> and count **leftward** when the **known number** is <u>positive</u> and **rightward** when it's <u>negative</u>. Then draw an **arrow** from the **X** value to the <u>**answer**</u>.

X + (- 3) = 4

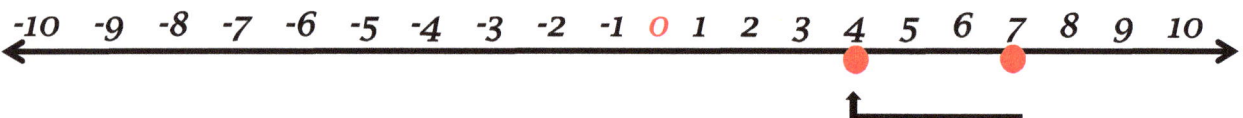

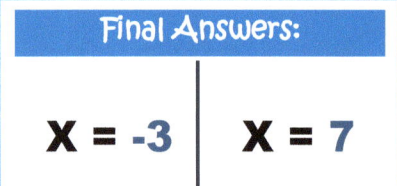

Final Answers:	
X = -3	X = 7

Check Your Answers:	
X + 8 = 5	X + (-3) = 4
-3 + 8 = 5 ✓	7 + (-3) = 4 ✓

Algebra Practice Problems – <u>Large Dots</u>

X + 7 = 9

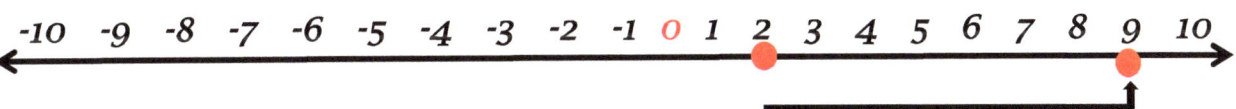

> To get the **X** value, start at the **<u>answer</u>** and count **leftward** when the **known number** is <u>positive</u> and **rightward** when it's <u>negative</u>. Then draw an **arrow** from the **X** value to the **<u>answer</u>**.

X + (- 7) = -2

Final Answers:	
X = 2	**X = 5**

Check Your Answers:	
X + 7 = 9	**X + (-7) = -2**
2 + 7 = 9 ✓	**5 + (-7) = -2** ✓

Algebra Practice Problems – <u>Large Dots</u>

X + (-9) = -9

To get the **X** value, start at the **<u>answer</u>** and count **leftward** when the **known number** is <u>positive</u>

and **rightward** when it's <u>negative</u>. Then draw an **arrow** from the **X** value to the **<u>answer</u>**.

X + (- 9) = -10

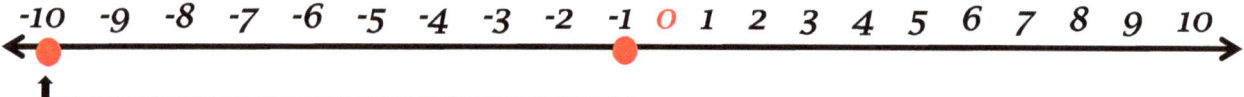

Final Answers:	
X = 0	**X = - 1**

Check Your Answers:	
X + (-9) = -9	**X + (-9) = -10**
0 + (-9) = -9 ✓	**-1 + (-9) = -10** ✓

Algebra Practice Problems – Large Dots

X + (-5) = -7

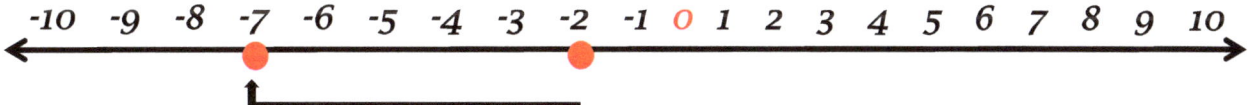

```
To get the X value, start at the answer and count leftward when the known number is positive
and rightward when it's negative. Then draw an arrow from the X value to the answer.
```

X + (- 3) = - 6

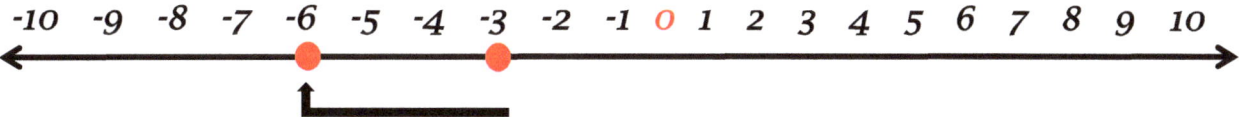

Final Answers:	
X = - 2	X = - 3

Check Your Answers:	
X + (-5) = -7	X + (-3) = -6
- 2 + (-5) = -7 ✓	- 3 + (-3) = -6 ✓

Algebra Practice Problems – Large Dots

X + (- 4) = 0

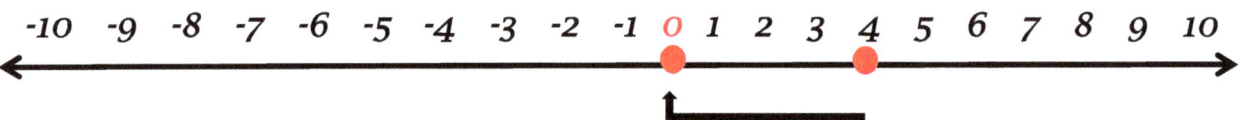

> To get the **X** value, start at the **answer** and count **leftward** when the **known number** is <u>positive</u> and **rightward** when it's <u>negative</u>. Then draw an **arrow** from the **X** value to the **answer**.

X + 9 = 2

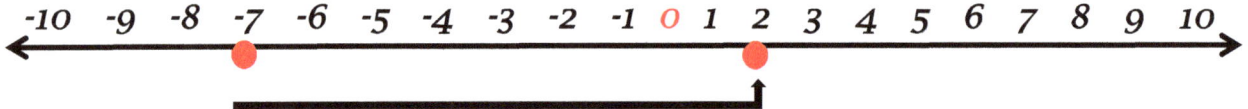

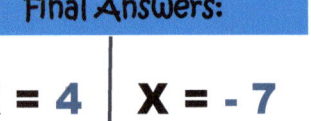

Final Answers:	
X = 4	X = - 7

Check Your Answers:	
X + -4 = 0	X + 9 = 2
4 + -4 = 0 ✓	-7 + 9 = 2 ✓

Algebra Practice Problems – <u>Large Dots</u>

X + 15 = 8

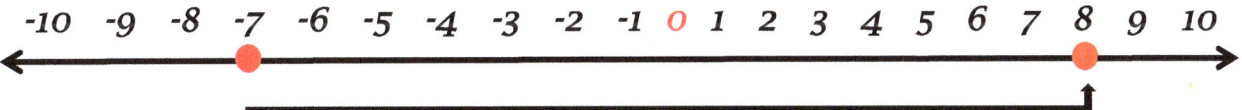

To get the **X** value, start at the **<u>answer</u>** and count **leftward** when the **known number** is <u>positive</u> and **rightward** when it's <u>negative</u>. Then draw an **arrow** from the **X** value to the **<u>answer</u>**.

X + 5 = - 4

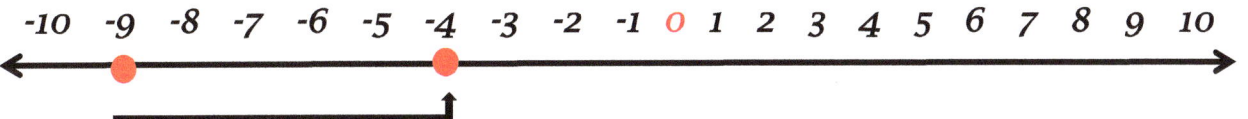

Final Answers:	
X = -7	**X = - 9**

Check Your Answers:	
X + 15 = 8	**X + 5 = - 4**
-7 + 15 = 8 ✓	**- 9 + 5 = - 4** ✓

Princess Sasha Saves Baby Dinosaurs

Algebra Practice Problems – <u>More Pictures</u>

X + (-2) = 3

X - (-2) = 7 $\longrightarrow$ **X** + 2 = 7

Final Answers:	
X = **5**	X = **5**

Check Your Answers:	
X + (-2) = 3	X - (-2) = 7
5 + (-2) = 3 ✓	**5** + 2 = 7 ✓

27

Algebra Practice Problems – <u>More Pictures</u>

$X + (-2) = 6$

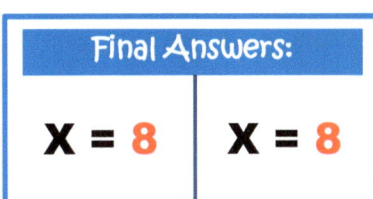

$X - (-2) = 10$ ⟶ $X + 2 = 10$

Final Answers:	
$X = 8$	$X = 8$

Check Your Answers:	
$X + (-2) = 6$	$X - (-2) = 10$
$8 + (-2) = 6$ ✓	$8 + 2 = 10$ ✓

Algebra Practice Problems - <u>More Pictures</u>

X + (- 4) = 5

X + ❌🦋 ❌🦋 ❌🦋 ❌🦋 = 🦋 🦋 🦋 🦋 🦋

X - (- 4) = 13 ⟶ **X** + 4 = 13

X + 🦋 🦋 🦋 🦋 **=** 🦋 🦋 🦋 🦋 🦋 🦋

🦋 🦋 🦋 🦋 🦋 🦋 🦋

Final Answers:	
X = **9**	**X** = **9**

Check Your Answers:	
X + (-4) = 5	X - (-4) = 13
9 + (-4) = 5 ✓	**9** + 4 = 13 ✓

Algebra Practice Problems – <u>More Pictures</u>

X + (- 5) = 1

X + ❌🐘 ❌🐘 ❌🐘 ❌🐘 ❌🐘 = 🐘

X - (- 5) = 11 ⟶ **X** + 5 = 11

X + 🐘🐘🐘🐘🐘 = 🐘🐘🐘🐘🐘

🐘🐘🐘🐘🐘🐘

Final Answers:	
X = 6	**X = 6**

Check Your Answers:	
X + (-5) = 1	X - (-5) = 11
6 + (-5) = 1 ✓	6 + 5 = 11 ✓

30

X + (- 3) = 15

X + **=**

Final Answer:

X = 18

Check Your Answer:

X + (- 3) = 15

18 + (- 3) = 15 ✓

18 - 3 = 15 ✓

31

Algebra Practice Problems – <u>More Pictures</u>

X - (- 7) = 7 → **X + 7 = 7**

X + 🦆🦆🦆🦆🦆🦆🦆 **=** 🦆🦆🦆 🦆🦆🦆 🦆

Final Answer:
X = 0

Check Your Answer:
X – (-7) = 7
0 + 7 = 7 ✓

Algebra Practice Problems – <u>More Pictures</u>

X - (- 4) = 12 ⟶ **X + 4 = 12**

X + 🐱🐱🐱🐱 = 🐱🐱🐱🐱
🐱🐱🐱🐱
🐱🐱🐱🐱

Final Answer:
X = 8

Check Your Answer:
X – (-4) = 12
8 + 4 = 12 ✓

33

Algebra Practice Problems – <u>More Pictures</u>

X - (- 3) = 19 ⟶ **X + 3 = 19**

X + 🦓 🦓 🦓 **=** 🦓 🦓 🦓 🦓 🦓

🦓 🦓 🦓 🦓 🦓

🦓 🦓 🦓 🦓 🦓

🦓 🦓 🦓 🦓

Final Answer:
X = 16

Check Your Answer:
X – (-3) = 19
16 + 3 = 19 ✓

Algebra Practice Problems – <u>More Pictures</u>

X - (- 1) = 15 ⟶ **X + 1 = 15**

X + 🐐 **=** (fifteen goats)

Final Answer:
X = 14

Check Your Answer:
X – (-1) = 15
14 + 1 = 15 ✓

Algebra Practice Problems – More Pictures

$X - (-3) = 3$ → $X + 3 = 3$

X + 🦖 🦖 🦖 = 🦖

🦖

🦖

<table>
<tr><td>**Final Answer:**</td></tr>
<tr><td>$X = 0$</td></tr>
</table>

<table>
<tr><td>**Check Your Answer:**</td></tr>
<tr><td>$X + 3 = 3$
$0 + 3 = 3$ ✓</td></tr>
</table>

About the Author

Dr. Courtney West was inspired to write *Princess Sasha Saves Baby Dinosaurs: Fun Algebra* after reading Dr. Keith Devlin's book, *The Math Gene*.

Dr. Devlin compared mathematical thinking to gossip. He suggested that Pre-K math education should be combined with stories that involve human interactions.

Princess Sasha Saves Baby Dinosaurs: Fun Algebra is the first release of a 12-book Algebra 1 series for children ages 4 and older. Upon completion of the series and its accompanying *Practice Problems* books, a 4-year-old child (after someone patiently reads the interactive books to him/her) should be able to solve all of the equations in Barron's *Painless Algebra* book, by Lynette Long, Ph.D.

Dr. West is the founder of www.sashasworld.biz. She has a Ph.D. in Mathematics Education and a J.D. emphasizing Juvenile Law from The University of Colorado - Boulder. Her B.A. degree in Journalism is from Rowan University of New Jersey. Dr. West resides in Centennial, Colorado.

www.ingramcontent.com/pod-product-compliance
Lightning Source LLC
Chambersburg PA
CBHW051104180526
45172CB00002B/771